*The botany of gin*

# *The* botany *of* gin

Chris Thorogood and
Simon Hiscock

# Contents

# Introduction

**ROOTED IN ANCIENT GREEK** herbal medicine, and established in the sixteenth century as a juniper-infused tincture to cure fevers, gin is defined by the plants with which it is made. A diverse assortment of plants from around the world have been used in its production over hundreds of years. Each combination of these botanicals (which include seeds, roots and bark) yields a unique flavour. Understanding the different types of formulation, and the plants used, is central to appreciating the complexities and subtleties of gin. Alongside the resurgence that gin is enjoying today, the use of a profusion of new botanical ingredients means that tasting gin can be likened to savouring a fine wine.

This book delves into the botany of gin from root to branch, and is garnished with illustrations depicting the plants that tell the story of this complex drink. It takes its inspiration from the gin botanicals that are grown in the University of Oxford Botanic Garden, plants that have themselves inspired the recipe for a new gin distilled in the city (see *Artisan gins*, p.15). The interpretive oil paintings by Chris Thorogood show the most prominent features of the plants, whether the leaves, flowers or fruits. Further information about which parts of the plants are used to make gin, and aspects of their general biology, are given in the descriptions.

## The origins of the use of plants in alcoholic beverages

Gin is a botanical beverage with a rich history like no other. It has its origins in herbal medicine in the Mediterranean.

Distillation technology (the process of purifying a liquid by heating and cooling) was pioneered by the Ancient Egyptians and Greeks, and it was used with various types of Mediterranean plants to flavour all sorts of alcoholic drinks including wines and tinctures (concentrated herbal extracts created by soaking botanicals in alcohol or vinegar). This technique is still essentially that used in the production of gin today.

The dominant form of vegetation in the Mediterranean is a dense, spiny thicket, which forms a dark green mantle on the red, rocky hillsides that are characteristic of this part of the world. This botanically rich community of plants, often known locally by the French names *maquis* and *garrigue*, contains a multitude of culinary herbs that are resinous and strongly aromatic. The plants' protective oils volatilize under the Mediterranean's searing sun into a heady bouquet of cistus, rosemary, thyme and sweet fennel. Plants such as dittany of Crete (*Origanum dictamnus*) and wormwood (*Artemisia* spp.), which still grow wild in the rocky crevices of Mediterranean mountainsides and gorges today, were well known in Classical Greece for their tonic and digestive properties. The Greek physician Hippocrates (fourth–fifth century BCE) used dittany of Crete as a treatment for various ailments and in poultices. He also reported that plants were macerated (soaked) to create aromatized wine, which throughout antiquity was known as 'Hippocratic wine'.[1] So began the use of herbs in the preparation of botanical tinctures. To this day, the distillation of aromatic plants is the main traditional method to extract essential oils, which are as important in flavouring spirits such as gin as they ever were.

The beginnings of gin

Gin derives its predominant flavour from the cones – usually known as 'berries' and referred to as such in this book – of juniper (*Juniperus communis*). Indeed to this day, juniper must be the dominant flavour in order for a drink to be classified legally as a gin. The production of gin started in seventeenth-century Holland, with the distillation of juniper berries immersed in water and alcohol. The first distilled spirit aromatized with juniper is attributed to a Dutch physician called François or Franz de le Böe, also known as Franciscus Sylvius, (1614–1672) who worked at Leiden University. A primitive form of gin was produced as an antidote for fever suffered by Dutch colonizers of the West Indies.

English soldiers serving in the Netherlands during the reign of Elizabeth I are said to have taken a dram of distilled spirit for 'Dutch courage' – a term that predates the first true, juniper-infused gin, and survives to this day. It is possible that upon returning to England these soldiers introduced the drinking habits acquired on the Continent to their compatriots.[2] Over the following three centuries, distilling grew steadily in popularity on the Continent, and gin was to change from a beverage for the privileged classes, much like brandy, to one that was mass-produced for the working classes. By the early eighteenth century, ale was being replaced by gin as a popular recreational alcoholic drink.

## The Gin Craze and beyond

In Britain, gin had emerged as an affordable, potent drink, which, importantly, could be produced locally – unlike wine, for example. It also had the perceived benefit of providing a market for the glut of surplus grain, and so its production was at first backed with enthusiasm by the government in support of farmers. Parliament did away with domestic distilling monopolies which had kept prices high. Now, anybody could pay the duties to set up a distillery and sell their produce, leading to a sharp fall in the price of spirits: gin became cheaper to drink than ale. And so began the Gin Craze.

In the eighteenth century, against a backdrop of overcrowding, poverty and social unrest,[3] London was pervaded by a growing black market of backstreet bars and gin pedlars. This development was fuelled by illicit stills, in which gin was distilled crudely and adulterated by all sorts of contaminants, some of them extremely poisonous – for example, methanol (a simple form of alcohol which is toxic if taken in large volumes). At the height of the Craze, London was described by some as a lawless, drunken city of staggering men and women, with an increasing crime rate. In fact, this turmoil was the result of a number of complex social and political issues. The 'new kind of drunkenness' was alarming because many of the consumers of this cheap and potent drink were the poor – if a violent and economically useless working class were to emerge, it would pose a threat to the country's very social fabric.[4] The artist William Hogarth (1697–1764) made two prints in support of the Gin Act of 1751: *Gin Lane* depicts the carnage and destitution brought about by the consumption of gin and other alcoholic drinks during the Craze, in contrast to *Beer Street*, which illustrates the benefits of British ale. With the Act, the government sought to curb the menacing, addictive gin-drinking epidemic and the perceived social decline with fines and prosecutions, and eventually, a prohibition on the sale of gin altogether. Nevertheless, illicit gin consumption continued for several years, until heavy taxes on large distillers, and poor grain harvests, together with a change in social conditions, eventually curbed the Craze.

A reduction in the duty on spirits during the budget of 1825 led to the resurgence of gin – no longer sold by backstreet distillers, but by 'gin palaces'. The first gin palaces were constructed in London at great cost, but unlike old-fashioned pubs and taverns had no seating or private sections. The gaudy Victorian gin palace provided an escape from the challenging conditions of nineteenth-century London and was a recapitulation, of sorts, of the Gin Craze a century before.[5]

In 1830 the Irish distiller Aeneas Coffey (1780–1839) introduced modifications to the recently developed column still to produce a lighter spirit with a higher alcohol content. His 'patent still' was taken up by gin distilleries across England and much improved the quality of their products. Gin as we know it today was born in the form of the first London Dry gins (see *Classical gins*, p.14). Finally, the respectability of gin was established when the 'gin and tonic' rose in popularity in the late nineteenth century (see *The gin and tonic*, p.16).

# How gin is made

The 'base' for gin is a rectified, neutral base spirit (a highly concentrated alcohol that has been purified through repeated distillation). The neutral spirit is typically derived from wheat, but can also be produced from barley, potato or sugar cane. In the seventeenth century the 'pot still' was used to produce alcoholic spirits in batches. Stills were traditionally constructed using copper, which removes unwanted sulphurous flavours, though today they often contain stainless steel and/or glass. The type of copper the still contains has an important impact on the character of the gin produced. The still distils the liquid by boiling and cooling it to condense the vapour. Aromatic compounds from the botanicals used become absorbed by the spirit in which they are immersed, or by the vapour which passes through them. With the advent of the column still (see *The Gin Craze and beyond*, p.9) in the 1820s, gin was produced on a continuous basis rather than in batches and the importance of 'clean' ethanol (alcohol) became better appreciated. This was the heyday of secret botanical formulas for gin.

Some craft distillers still make their own base spirit, although this can be expensive and time-consuming. This spirit can then be transformed to produce a gin by one of three principal methodologies: distilling (sometimes referred to as 'one-shot'), concentrating, or compounding:

## 1. Distilling

Distilled gins are produced by the distillation of a neutral spirit and water, together with juniper berries and other botanicals. The botanicals are often suspended within the still in baskets, through which the alcohol and water vapours pass to extract the flavours (a methodology known as 'vapour infusion'). The alcoholic distillate is then cut with water to the final 'proof' (the measure of ethanol in the beverage). This traditional method can be labour intensive, involving the meticulous preparation, weighing and measuring of botanicals for each batch, but is generally considered to be the best way to produce high-quality gins, and is favoured by most craft and artisan distillers.

## 2. Concentrating

Alternatively, gin can be produced by a simpler process in which juniper berries and other macerated botanicals are distilled with a small quantity of liquid to generate a highly concentrated distillate to which the neutral spirit and water are later added. The length of time the botanicals remain in the spirit varies from one distiller to another. Some distillers macerate various botanicals in the spirit and vapour infuse others, to tweak and fine-tune the strength of the different flavours.

## 3. Compounding

Finally, the compounding method is a very simple process, without distillation, by which the essential oils of juniper and other botanicals and flavourings, either natural or artificial, are added directly to the neutral spirit and water, then sieved out before bottling.

In addition to the conventional methods described above, some distillers are experimenting with new technologies, particularly with regard to temperature. Cold distilling is a method by which a vacuum reduces the temperature required for the ethanol to distil, which alters the flavour profile of the botanicals.

Regardless of the methodology, the research and development process for new gin formulations involves painstaking trial and error, using different combinations and quantities of plant. 'Flavour assays' are employed, in which exact quantities of botanical ingredients are modified meticulously according to their strength of flavour. These quantities are calculated by the weight of the botanical ingredient, and the amount of essential oil the material is established to contain. This is complicated by the fact that the oil content and flavour of the botanicals vary from year to year, according to the conditions under which the plants were grown, and with geographical origin. Therefore, specifications setting out, for example, the age of the plants and their oil and moisture content are used to ensure quality and consistency. For juniper berries, the specification may also include the size of the berries and the percentage that may be unripe. Distillers have different preferences for how dry, sharp or sweet the finished product should be, which will influence the preferred provenance of the juniper, and the age of the berries. Thus each distiller may have unique sources and specifications for their berries.

# Types of gin

Besides the different production methods for gin described above and the myriad possible botanical combinations, the finished product can be classified in many ways, for example by the location or scale of production, or by the amount of alcohol by volume (ABV). This is reflected in the assortment of different gins available to the shopper and served by many pubs and bars today. Broadly speaking, however, most gins can be defined as either 'classical' or 'contemporary'. Both styles have seen a recent rise in artisan production.

## Classical gins

London Dry Gin is a technical term meaning that the gin meets certain legal requirements relating to its method of production, rather than referring to its place of origin or flavour *per se*. These standards set out the ABV of the base spirit, and require that the dominant flavours must arise from distillation and no additional essences or colours can be added later. Several well-known 'heritage brands' are produced in this way, such as Gordon's London Dry Gin, Tanqueray London Dry Gin and Beefeater London Dry Gin. Bombay Dry Gin rekindled interest in gin following its launch in 1960, despite competition from other, more popular spirits at the time, such as vodka. Bombay Sapphire was an updated formulation of Bombay Dry Gin that was launched in the 1980s and remains popular to this day.[6] Each of these gins is characterized by traditional or 'classical' flavour profiles produced by the redistillation of the neutral spirit in the presence of various botanicals (see *How gin is made*, p.11). Juniper, of course, forms the basis of the formulation in all of these gins, and is noticeably dominant; it is often complemented by coriander seeds (technically fruits) and grains of paradise, and sometimes the peels of citrus. Numerous modern brands share this classical structure, often with a novel 'twist' such as one or two less conventional botanicals.

Dutch gin or 'genever' (derived from the Dutch word for juniper) is a classical-style gin produced in Holland that resembles the original gins of the seventeenth century. The malted spirit is based on wheat, barley, maize or rye, and the liquid double-distilled to yield a 'malt wine'. This is then redistilled with juniper berries and other botanicals, resulting in a heavily flavoured drink with an almond-like taste.

## Contemporary gins

In recent years, many variations on the basic flavour of gin have been developed, in which botanicals besides juniper are often more discernible, particularly those with citrus, herbal, floral or spicy notes. Some have a much more floral and fruity palate and are served with cucumber, for example (see *Garnishes*, p.102). Gin Mare (*mare* is Italian for 'sea') has a crisply herbal finish that is evocative of the Mediterranean, imparted by thyme, rosemary and basil in its formulation.

## Artisan gins

'Artisan' and 'hand-crafted' are widely used marketing terms which are not as strictly defined as, for example, alcohol content, sweetness and packaging. They generally relate to the scale and level of automation of gin production, including the batch size and process and the amount prepared by hand. Artisan or craft gins are seeing a rise in popularity with changing consumer awareness of provenance; indeed, some small distillers are now making a big impact on the gin landscape.

Physic Gin produced by the Oxford Artisan Distillery ('TOAD') is an artisan gin created in collaboration with the University of Oxford Botanic Garden. It is inspired by the original 'Physicke Garden', which was established in 1621 for the teaching of herbal medicine. Oxford Botanic Garden (as it is now known) – the oldest in Great Britain – was planted during the 1640s by its first keeper, Jacob Bobart the Elder. In 1648, Bobart created a catalogue of all the plants that he grew, now preserved as a treasured manuscript at the Garden. Many of the botanicals used in Physic Gin, which are harvested from the Garden itself, were listed in this catalogue: for example various citrus, which have been grown in Oxford for centuries, feature prominently in the formulation of the gin. Other botanicals used commonly in the making of gins can be seen growing in a special border in the Garden. Physic Gin is produced with stills repurposed from old copper steam engines, and the spirit is made with ancient 'heritage' grains. Other historic botanic gardens up and down the country are now also producing artisan gins linked to their plant collections.

# Gin in drinks

### The gin and tonic

Like gin itself, the origins of the gin and tonic are as a medicinal beverage. Tonic water is a bitter, carbonated drink in which botanical essences and, traditionally, quinine are dissolved. Quinine has been used since the seventeenth century as a prophylactic drug against malaria. It is an alkaloid isolated from the inner bark of cinchona trees, also known as 'Peruvian bark', from South America. *Cinchona* is a genus of trees and shrubs in the coffee family (Rubiaceae), native to the cloud forests of Bolivia, Colombia and Costa Rica, the medicinal properties of which are still not superseded by synthetics to this day.

In nineteenth-century India, the British authorities would ensure an ample supply of cinchona bark was available, and added to water, to protect their colonists. The most effective form was said to be derived from a particular species called the red cinchona (*Cinchona pubescens*). Due to its very bitter taste, it was mixed with sugar and soda to make it more palatable. Thus a primitive form of tonic water was born. This was at a time when gin – formerly associated with social decay (see *The Gin Craze and beyond*, p.9) – was enjoying a resurgence in respectability, and so the popular 'gin and tonic' was concocted. Today quinine is added in much lower quantities to tonic water. Now, a sharp hint of quinine is augmented by other bitter compounds and a range of botanicals, just like gin, culminating in a richly aromatic drink. Some of the botanicals used, for example juniper and citrus, mirror those used in gin, to give a refreshing taste. Mediterranean herbs are also used, as well as various blossoms such as elderflower, to impart floral notes.

### Gin cocktails

Some of the most popular cocktails are made with a base of gin, owing to its subtle, aromatic flavours; just a few of the better-known ones are described here. The Pink Gin is believed to have origins in the Royal Navy, in the nineteenth century. It contains red bitters (various bitter-tasting botanical extracts including citrus), which give the drink its characteristic rose-coloured finish, and is often served with ice, water and lemon peel. The Gin Martini is another nineteenth-century classic, made with vermouth (a flavoured wine), also enhanced with a twist of lemon. The Negroni is a popular twentieth-century Italian cocktail that also contains gin and vermouth, along with the liqueur Campari (see *Other botanical spirits*, below). It is widely served with orange peel and taken as an apéritif; in some parts of Sicily, it is served with a slice of blood orange (*Citrus* x *sinensis*), a particular kind of orange with dark red flesh. The Gin Sling or Singapore Sling is another regional twentieth-century classic that was made traditionally with cherry brandy and various fruit juices, and today, often with grenadine.

# Other botanical spirits

There exists a plethora of spirits infused with botanicals besides gin, the most famous being the vermouths, bitters and liqueurs. They too evolved from the use of aromatic plants in wines and spirits that dates back to antiquity (see *The origins of the use of plants in alcoholic beverages*, p.7), and many have their beginnings in traditional remedies. For example, aromatized wines infused with macerated bark, stems and botanicals – gentian and rhubarb in particular – were used in nineteenth-century Italy as a cure for fevers.

Vermouths are a form of aromatized wine that date back to the seventeenth century. The name is derived from *vermut* (German for wormwood), or 'wermut-wine', which was produced in Bavaria; 'vermouth' is modified from the French spelling.[7] Various botanicals are used in vermouth besides wormwood itself (*Artemisia absinthium*), which is the dominant constituent, including plants often found in the production of gin such as cloves, coriander, angelica, cinnamon and herbs.

Bitters, among the best-known of which is Campari, also date back to antiquity, and comprise a range of spirits flavoured with bitter, sour or bittersweet botanicals. Many were developed originally as herbal medicines but are now taken as recreational drinks, for example, as apéritifs. Botanicals typically used in their preparation include aromatic barks, plant roots and herbs or vegetables such as gentian, rhubarb and artichoke. Botanicals commonly used in bitters that are also used in the production of gin include the peel of various citrus fruits.

Many hundreds of liqueurs have been developed from botanical blends. Among the better-known are the aniseed-flavoured liqueurs, of which absinthe is probably the most famous. Absinthe was originally produced by the maceration of aniseed (*Pimpinella anisum*) and wormwood alongside other botanicals. Other liqueurs in this family include ouzo, raki and sambuca, which share similar, aniseed-dominated flavour profiles.

# The botanicals

To appreciate the complexity of gin, it is important to have an understanding of the botanicals that define the drink. Besides the *sine qua non* that is juniper, a multitude of aromatic plants, including seeds, roots and bark are also used in the production of gin. The different combinations of botanicals, their relative quantities and provenance all combine to make each formulation of gin unique.

## Aromatic plants

Aromatic plants are often richly scented when crushed because they contain essential oils, complex concentrated oils that contain aromatic and volatile (freely evaporated) compounds. These are produced in small quantities in various parts of the plant – including the leaves, stems, bark, flowers and seeds – and can be labour-intensive and expensive to extract. They are generally processed either by distillation in the presence of steam or by cold-pressing – crushing the plant material to release the oils, which are then collected and separated from the aqueous (watery) layer.

## Types of botanical

Although the pine-like flavour of juniper and, to a lesser extent, the mellow, spicy notes of coriander and angelica are the primary botanical flavours of gin, many other plants have been used historically, and are used today, in its preparation. Indeed a fine gin may contain ten or more different botanicals. The different combinations of botanicals culminate in all sorts of subtly different 'mouth-feel' and flavour profiles, ranging from floral, herby and fruity, to spicy, earthy and 'rooty'. 'Botanical tasting wheels'[8] have been developed to chart these complex and subtle flavour profiles and ascribe them to numerous different plants, guiding the taster to the relevant descriptors (for example 'woody, rich and creamy' and 'clean, crisp and ripe').

There are many ways in which the rich assortment of botanicals used in gin can be grouped. They are categorized primarily according to the character they confer on the finished product: bitter, such as gentian root or wormwood; aromatic, such as star anise or citrus; or spicy, such as clove, cinnamon or nutmeg. In this book, we have organized them into five groups which are based broadly on these flavours, and on the combinations in which they are typically used. They are not grouped by their botanical relatedness: for example, coconuts and baobabs have little connection other than that they are both tropical fruits used in contemporary gin formulations, which justifies their appearance in the same section. In addition, to avoid confusion, more familiar terms (berries, seeds and so on) are used in favour of precise botanical terms. Therefore we refer to juniper 'berries', which are in fact more akin to miniature pinecones botanically speaking, while cumin 'seeds' are in fact fruits, known technically as mericarps.

Although the list of possible botanicals used in gin is seemingly endless, the use of some is restricted because of their toxicity. Generally, however, the quantities of plant material in the finished product are only at trace level. Indeed, the quantity of gin that would need to be consumed for most botanicals to be considered dangerous would be toxic in and of itself!

# Fruits and berries

While its 'berries' are not technically fruits (see above), juniper must be given first place in our list because it is the predominant flavouring component of all gins. Citrus are an important class of fruits used in the preparation of gin, and as a popular garnish. The aromatic oil-rich peel of lemon, sweet orange and bitter orange, in particular, are all common ingredients, though limes and grapefruits are also used occasionally. Unsurprisingly, all these fruits confer fresh citrus notes to the finished product. Almost all the citrus described here are hybrid cultivars (that is, not naturally occurring species). The progenitors of cultivated citrus crops are native to Asia.

# Common Juniper

SCIENTIFIC NAME: *Juniperus communis*
FAMILY: Cypress family (Cupressaceae)

DESCRIPTION: Juniper is an evergreen prickly bush, or tree, related to cypresses and pines. The leaves are needle-like or scale-like. The dull, blue-black 'berries' are in fact female cones (like those of pine trees), in which fruit-like scales coalesce to form a fleshy coating.

DISTRIBUTION: Widespread across the Northern Hemisphere, from the Arctic to the mountains of Central America, Europe and Asia, with remote populations in North Africa.

USE IN GIN: Juniper is the foundation of all gin formulations, and must be the dominant flavour in order for a drink to be classified legally as a gin. The berries are rich in essential oils that confer a fresh green, fragrant and pine-like, and often slightly fruity flavour with a peppery finish. The precise flavour of the berries varies with the provenance of the plant. The finest juniper berries are to be found on the mountain slopes of south-central Europe, in Macedonia and Italy. Distillers' preference for either a dry, sharp or sweet formulation influences the provenance and the age of the berries they select.

# Sweet orange

SCIENTIFIC NAME: *Citrus* x *sinensis*
FAMILY: Rue family (Rutaceae)

DESCRIPTION: The sweet orange is a small, shallow-rooted citrus tree with spiny branches that bear the familiar edible fruit (botanically classified as a type of berry called a hesperidium that has a leathery rind with oil glands). The sweet orange group includes some of the most popular citrus fruits and many forms are cultivated, primarily for their juice, but also for seed oil, which is used in cosmetics. The blood orange is an unusual variant with particularly dark or red-streaked flesh that has been grown in the Mediterranean (especially in Spain and Italy) since the eighteenth century.

DISTRIBUTION: Sweet oranges are cultivated across the world in frost-free climates.

USE IN GIN: Sweet orange gives gin a zesty, refreshing taste.

# Lemon

SCIENTIFIC NAME: *Citrus* x *limon*
FAMILY: Rue family (Rutaceae)

DESCRIPTION: Lemon trees are hybrid cultivars bred for their versatile fruits. Typically they are spiny evergreens with leathery, aromatic leaves and fragrant white or pink-tinged flowers. After the synthesis of the bitter orange (*Citrus* x *aurantium*), crosses between oranges and a citrus species called the citron (*Citrus medica*), a plant with particularly large fruits native to northeast India, gave rise to the lemon. Owing to its distinctive taste, the lemon has become a key ingredient in many drinks and foods. It also has many non-culinary purposes.

DISTRIBUTION: Cultivated widely throughout warmer climates.

USE IN GIN: Lemon is the most popular citrus fruit to be used in gin and its flavour remains identifiable after the distilling process. It is also, of course, a popular garnish.

# Lime

SCIENTIFIC NAME: *Citrus* x *aurantiifolia*
FAMILY: Rue family (Rutaceae)

DESCRIPTION: 'Lime' is a name used to describe a whole range of fruits (some of which are unrelated), including sunrise lime, Australian finger lime, Persian lime, Key lime, leech lime (see p. 80), Rangpur lime, sweet lime and so on. The plant illustrated is the common artificial hybrid derived, at least in part, from the pomelo (*Citrus maxima*) – the largest citrus fruit, native to South and Southeast Asia, with an appearance somewhat similar to a grapefruit. The lime is a small, thorny tree with leaves rather like those of an orange tree (the scientific name *aurantiifolia* refers to the resemblance to the leaves of the bitter orange, *Citrus* x *aurantium*).

DISTRIBUTION: Cultivated widely throughout frost-free climates.

USE IN GIN: Lime is a popular garnish, of course, and many gins include lime as a botanical. Lime confers a lively, slightly bitter 'green' taste.

# Chinese bitter orange

SCIENTIFIC NAME: *Citrus trifoliata* (syn. *Poncirus trifoliata*)
FAMILY: Rue family (Rutaceae)

DESCRIPTION: Chinese bitter orange has robust, finger-long thorns and white flowers. The fruits are greenish-yellow and resemble a small orange, with a surface texture like that of a peach. It is an unusually cold-hardy citrus that can even tolerate frosts and snow, and other, tenderer species of citrus are often grafted onto the rootstock of this tree.

DISTRIBUTION: Northern China and Korea.

USE IN GIN: Chinese bitter orange is an unusual ingredient, not used widely in gin. It confers a sharp, citrusy note.

# Grapefruit

SCIENTIFIC NAME: *Citrus* x *paradisi*
FAMILY: Rue family (Rutaceae)

DESCRIPTION: The grapefruit is a small tree that grows to about 6 m high with glossy, dark green leaves, white, sweet-smelling blossom, and large fruits which are borne in clusters (apparently vaguely reminiscent of giant grapes, hence its name). Like oranges, grapefruits are derived from artificial crosses between mandarins (originating in China) and pomelos (from South and Southeast Asia). The different cultivars yield fruits with white, pink, or red flesh of varying sweetness (typically the redder fruits are sweeter).

DISTRIBUTION: Cultivated widely in frost-free climates.

USE IN GIN: Grapefruit is used in gin less often than other citrus because of its dominant flavour but it features in one or two brands as a 'signature botanical'. It can be used as a garnish in such gins, as well as in other gins in which citrus notes are not too intense already.

# Mandarin orange

SCIENTIFIC NAME: *Citrus reticulata*
FAMILY: Rue family (Rutaceae)

DESCRIPTION: Like many citrus, the mandarin is a small, thorny tree that grows to about 7–8 m high. The small, rather flattened fruit has a thin peel and segments that separate freely. It is generally considered to have sweeter flesh than the orange (*C.* x *sinensis*), which is derived from it. The oil of mandarin orange is used in the cosmetics industry.

DISTRIBUTION: The mandarin is not a particularly cold-tolerant tree so is restricted, on the whole, to warmer (especially tropical and subtropical) climates.

USE IN GIN: Mandarin is used in Masons Gin and its peel features, alongside several other citrus, in Physic Gin, the gin inspired by the Oxford Botanic Garden.

# Cucumber

SCIENTIFIC NAME: *Cucumis sativus*
FAMILY: Gourd family (Cucurbitaceae)

DESCRIPTION: The cucumber is a vigorous creeping vine that is related to pumpkins and gourds. Its stems are often trained up canes and trellises to which it clings with coiled tendrils. The rather large, yellow flowers are typically unisexual (either male or female), meaning that cross-pollination by insects or by hand is required for the plants to bear cucumbers. The fruit itself (often prepared as a vegetable) is a watery, modified berry with a thin, rather hard skin.

DISTRIBUTION: Originally from southern Asia, the cucumber is cultivated on every continent.

USE IN GIN: Hendrick's Gin is famously served with cucumber, which is also one of the botanicals used in its formulation.

# Tropical fruits

The revival of gin has seen a vast expansion in the range of botanicals used to produce 'contemporary' gins. Recently, distillers have experimented with a variety of tropical fruits such as coconuts, baobab and longan. These unconventional botanicals reflect changes in consumer tastes, and their use is an important chapter in the story of the evolution of gin.

# Longan

SCIENTIFIC NAME: *Dimocarpus longan*
FAMILY: Soapberry family (Sapindaceae)

DESCRIPTION: The longan is a large tropical tree that produces yellow flowers, followed by fruits borne in drooping clusters. This peculiar fruit is said to look like an eyeball when shelled, owing to its black seed surrounded by translucent white flesh (hence its other popular name, 'dragon's eye'). The fruit is rich in vitamin C and has a long history of traditional use.

DISTRIBUTION: Native to South Asia.

USE IN GIN: The longan has a sweet, nutty flavour that is similar to that of the papaya or lychee, the latter belonging in the same plant family. Used recently in the formulation of one or two tropical-branded gins, it is also served occasionally with gin in cocktails as a garnish.

# Coconut

SCIENTIFIC NAME: *Cocos nucifera*
FAMILY: Palm family (Arecaceae)

DESCRIPTION: The coconut is a type of palm, and the only living member of the genus *Cocos*. It is a tall tree, growing to about 30 m high, with a smooth trunk. The fruit, which is fleshy inside, is known technically as a drupe, and weighs about 1.5 kg when mature. The flesh and the milk that are extracted from the versatile fruit are used in many cosmetic and food products; the fibrous outer coating of the fruit is used in furnishings.

DISTRIBUTION: Coastal areas throughout the tropics.

USE IN GIN: Coconut confers a delicate creaminess, distinct from the other botanicals described here. It is used to create a signature taste in tropical-themed gins.

# Baobab

SCIENTIFIC NAME: *Adansonia digitata*
FAMILY: Mallow family (Malvaceae)

DESCRIPTION: This extraordinary species is the most widespread of a group of trees known collectively as the baobabs. All are long-lived pachycauls – distinctive plants characterized by their distended trunks, which store water and are an adaption to survival under extremely dry conditions. During the dry season they shed their leaves. The flowers are whitish, large and heavy, growing to about 12 cm across, and are pollinated by fruit bats. The large egg-shaped fruits are initially green and later turn brown and harden.

DISTRIBUTION: Africa.

USE IN GIN: The fruit of the baobab is said to confer a citrus note to gin. Its use is the perfect example of distillers becoming more adventurous in their choice of botanicals, particularly those that denote provenance from a particular region.

# Vanilla

SCIENTIFIC NAME: *Vanilla planifolia*
FAMILY: Orchid family (Orchidaceae)

DESCRIPTION: This beautiful orchid is the source of vanilla –
one of the world's most popular flavourings. It is a vine-like
plant with fleshy roots, broad leaves and large greenish-
yellow flowers. The fruits are elongated pods that grow to
about 25 cm long and contain tiny, dust-like seeds which
become exposed when the ripe pods split open. The fruits
are harvested and fermented to produce commercial vanilla.
Owing to its difficulty of cultivation and small yields, like the
saffron crocus (see p.53), this botanical is expensive.

DISTRIBUTION: Native to Mexico and Central America;
introduced locally to other tropical regions.

USE IN GIN: Unlike many of the tropical plants described
in this book, creamy-flavoured vanilla is used widely in gin
formulations, typically as a subsidiary to enhance or balance
the flavours of other botanicals.

# Dried fruits, seeds and spices

Numerous fruiting pods, seeds and spices – many of which are ground – are used in the formulation of gin. Coriander seeds are one of the most widely used ingredients besides juniper berries. This flavoursome botanical contains linalool, a naturally occurring compound in aromatic plants, which gives gin a mellow, spicy aroma and taste. Its relative angelica is also used occasionally. Besides the dried fruits, seeds and spices described below, numerous others are used to give gin a subtle and complex flavour profile, including cubeb, which imparts a spicy, often slightly peppery or menthol-like note, and ground almonds, which give a nutty, marzipan-like taste.

# Saffron crocus

SCIENTIFIC NAME: *Crocus sativus*
FAMILY: Iris family (Iridaceae)

DESCRIPTION: This beautiful crocus is the source of saffron, a spice which is derived from the flower's three bright red stigmas and yellowish styles, and is among the world's most expensive spices. The plant has been cultivated since antiquity for traditional herbal medicine. Many biologically active constituents have been isolated from the saffron crocus, and extensive research has been undertaken into its medicinal properties. For example, extracts from the plant have been trialled in the treatment of mild to moderate depression.

DISTRIBUTION: The cultivated form is derived from a wild species that originates in the eastern Mediterranean. It is planted widely in Iran, Greece and India.

USE IN GIN: Saffron has been used in several commercial gin formulations. It infuses the spirit with a beautiful golden amber glow and a subtle and complex earthy-floral flavour.

# Allspice

SCIENTIFIC NAME: *Pimenta dioica*
FAMILY: Myrtle family (Myrtaceae)

DESCRIPTION: Also known as pimento, allspice is derived from the dried fruits of an evergreen tree that grows to about 20 m high. The fruits, which grow to about 10 mm across, are covered in little glands. They are picked green and dried in the sun, yielding the familiar brown, wrinkled peppercorn-like grains. These are used widely in cuisine, particularly in the plant's native Caribbean, and in the Middle East; allspice is a popular ingredient in cakes and baking.

DISTRIBUTION: Native to Central America and cultivated extensively in Jamaica; also grown widely in warmer climates in the Western Hemisphere.

USE IN GIN: The grains contain highly aromatic oils which convey a rich, peppery and spicy flavour and a somewhat oriental fragrance that is reminiscent of nutmeg, cloves and cinnamon. Allspice is a commonly used botanical in gin formulations.

# Cocoa

SCIENTIFIC NAME: *Theobroma cacao*
FAMILY: Mallow family (Malvaceae)

DESCRIPTION: This small evergreen tree is the source of cocoa beans – seeds which are used globally to produce chocolate. The flowers and fruits of cocoa sprout directly from the main trunk and branches of the tree – a characteristic known botanically as cauliflory. The tree was exploited long before chocolate was invented – the edible properties of the fruit were discovered over 2,000 years ago by the people of Central America living deep in the tropical rainforests, and the plant was used for various ceremonial, medicinal and culinary purposes. Today cocoa is used for a variety of pharmaceutical and cosmetic products besides being a significant cash crop for the production of confectionery.

DISTRIBUTION: Native to Mexico and Central and South America; introduced locally to other tropical regions across Africa and Asia.

USE IN GIN: Not a traditional botanical, cocoa nibs (fragments of fermented, roasted and crushed cocoa beans) have been used as a novelty ingredient in a few gin formulations, particularly alongside nuts and vanilla. In recent years, however, cocoa has appeared as a signature botanical in several widely available gins.

# Nutmeg

SCIENTIFIC NAME: *Myristica fragrans*
FAMILY: Mace family (Myristicaceae)

DESCRIPTION: Nutmeg is an evergreen tree that grows to about 15 m high, with dark green leaves. The seeds have a long history of use in medicine, as a spice and for preservation. The spice is still used widely today in various cosmetics and in food, although it can be harmful in large quantities. Mace is the lacy red structure (known botanically as an aril) that envelops the shiny seed, splitting when ripe; the mace is also used as a flavouring agent.

DISTRIBUTION: Native to Indonesia and grown widely in the tropics.

USE IN GIN: Nutmeg gives gin a warm, spicy, earthy taste. It contributes a 'woody' structure to the flavour profile and is used widely by distillers.

# Szechuan pepper

SCIENTIFIC NAME: *Zanthoxylum simulans*
FAMILY: Rue family (Rutaceae)

DESCRIPTION: Szechuan pepper is produced from several shrubby species in the genus *Zanthoxylum*. *Zanthoxylum simulans* is a spreading shrub with ash-like, pinnate (divided) leaves comprising about ten pairs of leaflets and sprays of yellowish-green flowers. The fruits, known botanically as follicles (dry fruits that split along one side) are small, about 3–4 mm across, and reddish-brown; when ripe, they split to release black seeds. This species grows in thickets in ravines and woods and on hillsides, and is cultivated widely as a condiment.

DISTRIBUTION: Native to parts of East Asia.

USE IN GIN: Not used widely but now reported in several gins, Szechuan peppercorns give a peppery heat and a warm finish that is best appreciated if the spirit is taken neat.

# Cardamom

SCIENTIFIC NAME: *Elettaria cardamomum*
FAMILY: Ginger family (Zingiberaceae)

DESCRIPTION: This relative of the ginger produces dark green leaves in double ranks. The whitish-lilac flowers appear at the base of the plant in loose clusters, and are followed by three-angled yellowish pods, which are harvested. These dried fruits and numerous black seeds have been used as a culinary spice and in traditional herbal medicines since the fourth century BCE.

DISTRIBUTION: Native to southern India and grown throughout the tropics.

USE IN GIN: Cardamom is an expensive spice, but only a little is needed to impart its complex sweet flavour, which is both leafy and vaguely like that of citrus.

# Grains of paradise

SCIENTIFIC NAME: *Aframomum melegueta*
FAMILY: Ginger family (Zingiberaceae)

DESCRIPTION: The plant from which these seeds are harvested is a herbaceous perennial with shiny red fruiting pods that grow to about 7 cm long and contain numerous reddish-brown seeds. These seeds are ground and used to flavour soups, stews and grilled meat, particularly in West African cuisine.

DISTRIBUTION: Native to tropical Africa.

USE IN GIN: Grains of paradise impart a peppery-citrus flavour, with a hint of gingery warmth.

# Black pepper

SCIENTIFIC NAME: *Piper nigrum*
FAMILY: Pepper family (Piperaceae)

DESCRIPTION: The plant is a climber that grows up to 10 m tall. The berry-like fruits are picked early for green pepper, later for black pepper, and when fully ripe, for white pepper (for which the outer layer is removed). These peppercorns are among the world's most widely used spices, flavouring cuisine of all kinds. Black pepper is also used in traditional herbal medicine, especially for digestive ailments.

DISTRIBUTION: Native to mountains in India but cultivated widely across the tropics.

USE IN GIN: Black peppercorns are a popular spicy, warming constituent in the formulation of gin.

# Cumin and coriander

SCIENTIFIC NAME: *Cuminum cyminum*, *Coriandrum sativum*
FAMILY: Carrot family (Apiaceae)

DESCRIPTION: Both cumin and coriander are slender, aromatic annuals with small white flowers. Their 'seeds' (which botanically speaking are tiny fruits called mericarps) are similar, and are dried and used widely as spices. Cumin was among the earliest herbs to be cultivated across Asia, Africa and Europe and has been used extensively in cooking and traditional medicine. Fresh coriander leaves are employed as a culinary herb and the dried seeds can flavour confectionery, bread or curry powder, and are used in cosmetics.

DISTRIBUTION: Cumin and coriander are widespread across central Asia and the Middle East and cultivated throughout warmer parts of the world.

USE IN GIN: Both spices are used widely in gin, especially coriander. Cumin gives the spirit an intense earthy and nutty flavour, while coriander confers warm, spicy notes.

# Opium poppy

SCIENTIFIC NAME: *Papaver somniferum*
FAMILY: Poppy family (Papaveraceae)

DESCRIPTION: This annual herb grows to about 1 m high and has grey-blue leaves. The flowers are variably white, mauve or red with darker markings. The fruits, which contain numerous seeds, are fluted and have caps with about 12 radiating, spoke-like ridges. Opium poppies have been cultivated since antiquity. They are now grown as garden ornamentals and as a source of poppy seeds. Many varieties exist and not all of them yield opium.

DISTRIBUTION: Probably native to the eastern Mediterranean and Middle East but cultivated widely.

USE IN GIN: Poppy seeds are not used commonly as botanicals but can impart a subtle, nutty, earthy taste.

# Star anise

SCIENTIFIC NAME: *Illicium verum*
FAMILY: Star anise family (Schisandraceae)

DESCRIPTION: Star anise is an evergreen tree that grows to about 15 m tall, with distinctive star-shaped fruits. It is quite distinct from anise (*Pimpinella anisum*), which is in the carrot family (Apiaceae), although both plants yield an essential oil containing anethole, which is used for flavouring confectionery. Star anise has been cultivated since about 2,000 BCE, and has a long history of use in Chinese traditional medicine for many ailments. The seeds are also chewed to aid digestion.

DISTRIBUTION: Probably native to southern China and northeast Vietnam; cultivated widely in warmer climates.

USE IN GIN: Star anise is an aromatic spice with an intense sweet, aniseed-like flavour. It is used in the production of several alcoholic beverages – sambuca, pastis and some formulations of absinthe – as well as in gin. It imparts a similar flavour to anise and has become a popular botanical. The fruits can even be added to a gin and tonic as a beautiful garnish.

# Angelica

SCIENTIFIC NAME: *Angelica archangelica*
FAMILY: Carrot family (Apiaceae)

DESCRIPTION: Angelica is a tall, aromatic herb that has been cultivated since antiquity for medicinal purposes, as a flavouring agent and as a vegetable. The individual whitish or greenish flowers are small, about 4 mm across, and are borne in large heads up to about 15 cm across. The candied stalks of angelica are used to decorate cakes and puddings, and in some countries the plant is still eaten as a vegetable.

DISTRIBUTION: Native to northern temperate regions.

USE IN GIN: Angelica is used to flavour liqueurs such as Bénédictine. Recently, the roots and seeds have also been used in the formulation of gin. They impart a mild, earthy aniseed-like flavour.

# Leaves and stems

The leaves and stems of numerous plants are used as
botanicals in gin. Sprigs of Mediterranean herbs such as sage,
rosemary and thyme are particularly popular, and bestow
a herbal, often pine-like finish. The leaves of the leech or
makrut lime, commonly found in Asian cuisine, are also used
occasionally in addition to the citrus fruits described above.

# Bay laurel

SCIENTIFIC NAME: *Laurus nobilis*
FAMILY: Laurel family (Lauraceae)

DESCRIPTION: The bay laurel is an evergreen tree that grows to about 10 m high, with leathery aromatic leaves which are a popular ingredient in cooking, particularly in Mediterranean cuisine. The leaves are added to stews and sauces and typically removed prior to serving. The plant is a relic of ancient laurel forests that once grew extensively across the Mediterranean Basin.

DISTRIBUTION: Native to the Mediterranean and grown widely in gardens in temperate climates around the world.

USE IN GIN: Bay laurel infuses spirits with an opulent green, Mediterranean flavour. It is used in the formulation of several brands of gin.

# Leech or makrut lime

SCIENTIFIC NAME: *Citrus hystrix*
FAMILY: Rue family (Rutaceae)

DESCRIPTION: Leech lime (historically known as kaffir lime) is a thorny tree with distinctive hour-glass shaped or 'double' leaves – the lower 'leaf' is in fact an expanded leaf stalk (petiole). The fruit is green and distinctly dimpled and warty. Also known as the makrut lime, its citronella-rich leaves emit an intense aroma when crushed and are popular in Thai cuisine. As well as being used for cooking, the fruit has been used to repel leeches (hence its name) and for washing hair.

DISTRIBUTION: Apparently wild in parts of Southeast Asia; also cultivated widely for culinary use.

USE IN GIN: Like lime peel, leech lime leaves give a zingy, fresh taste and confer complexity to the citrus flavour profile of a gin.

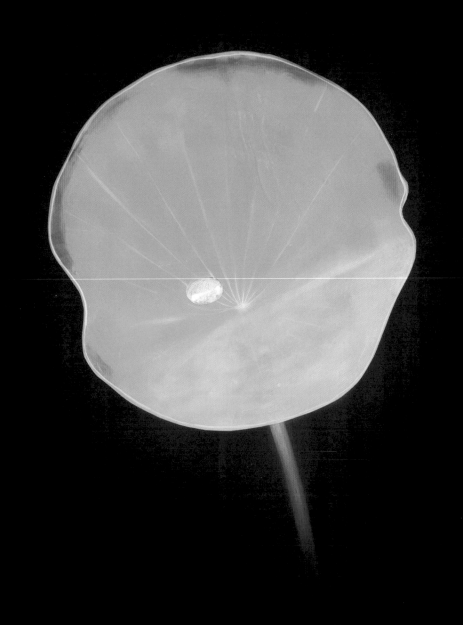

# Sacred lotus

SCIENTIFIC NAME: *Nelumbo nucifera*
FAMILY: Lotus family (Nelumbonaceae)

DESCRIPTION: The roots of this exquisite aquatic marginal plant anchor to the mud of river bottoms and its leaves and flowers emerge above the surface of the water on long stems. It floats unmarked above the muddy river waters beneath, and so is regarded as sacred by Hindus and Buddhists. It is cultivated for its showy flowers, which are used as religious ornaments, and is a cash crop for its edible rhizomes, which are eaten as fresh vegetables, and its seeds, which are used in desserts and for medicine.

DISTRIBUTION: Subtropical and tropical Asia.

USE IN GIN: Although far from conventional ingredients, lotus leaves, flowers and roots have all appeared in craft gin formulations recently as 'novelty botanicals' to give a leafy or floral freshness, or earthiness, to the drink, depending on the parts used.

# Absinthe wormwood

SCIENTIFIC NAME: *Artemisia absinthium*
FAMILY: Daisy family (Asteraceae)

DESCRIPTION: This plant is a woody-based shrub with silvery-blue, dissected leaves that emit a pungent aroma when crushed. It has a very long history of use in infusions, as a traditional medicine and as a stimulant. The plant is now best known for being a key ingredient in the formulation of the spirit absinthe, and it is used in some other spirits including bitters and vermouth. It has featured prominently in English literature for centuries.

DISTRIBUTION: Widespread across northern temperate regions.

USE IN GIN: Absinthe wormwood is not used commonly in gin. It infuses the drink with a bitter, green, 'forest floor' flavour.

# Lavender

SCIENTIFIC NAME: *Lavandula angustifolia*
FAMILY: Mint family (Lamiaceae)

DESCRIPTION: Lavender has a long and rich history of cultivation. Its flowers are strongly fragrant and yield large quantities of nectar from which high-quality honey can be produced. Today, the plant is grown mainly for the production of essential oils which are widely used in perfumery, and as fragrances for cosmetics and bath products. Lavender yields an essential oil with sweet overtones and is used commonly in balms, salves, perfumes, cosmetics and topical applications. The plant is also widely grown as a garden plant, for cut dried flowers and even for cooking.

DISTRIBUTION: Native to Mediterranean Europe and planted widely elsewhere.

USE IN GIN: Used in moderation, lavender can give gin a pleasant floral finish that complements other herbs and spices.

# Lemongrass

SCIENTIFIC NAME: *Cymbopogon citratus*
FAMILY: Grass family (Poaceae)

DESCRIPTION: Lemongrass is a tall, clump-forming grass that has been cultivated for centuries for its pungent citrus-scented foliage. The leaves can be dried and brewed to make tea, or used in cooking – they are a particularly popular ingredient in chicken dishes in the Philippines and Indonesia. Aromatic oils derived from the plant are also used in perfumery.

DISTRIBUTION: Native to Southeast Asia but grown widely throughout the tropics.

USE IN GIN: Lemongrass has a distinct, fresh lemon flavour that lacks the tartness of the lemon fruit. It is a popular ingredient in the formulation of citrus-packed gins.

# Roots, rhizomes and bark

The roots, rhizomes (which are, botanically speaking, underground, root-like stems) and bark of certain botanicals give structure to the flavour profile of a gin. One widely used botanical is dried orris root (technically a rhizome), which is derived from bearded irises (*Iris germanica* and *I. pallida*). The rolled bark of cassia, also known as Chinese cinnamon, and other forms of cinnamon contain the essential oil cinnamaldehyde, which gives gin a subtle spicy note. Finally, the root of liquorice imparts a similar sweet flavour to the unrelated plants star anise and fennel. The former is also often used in gin (see p.73).

# Sweet iris

SCIENTIFIC NAME: *Iris pallida*
FAMILY: Iris family (Iridaceae)

DESCRIPTION: This beautiful plant has straight, sword-shaped leaves and purple flowers which have a sweet, violet-like fragrance. The iris produces a rhizome that grows along the surface of the ground, somewhat resembling root ginger: this is the part, known as orris root, that is used in gin.

DISTRIBUTION: Native to Italy and the Balkans; grown widely across Europe.

USE IN GIN: The sweet iris has long been associated with the production of gin. Its rhizome is dried – often for several years – and ground into a powder. It imparts a sweet, floral, violet-like flavour and is one of the principle botanical constituents of gin.

# Ginger

SCIENTIFIC NAME: *Zingiber officinale*
FAMILY: Ginger family (Zingiberaceae)

DESCRIPTION: The leaves of ginger grow to over 1 m high, and the flowers are borne in unusual cone-like structures. The rhizome was one of the first oriental spices to arrive in Europe and is now used widely in cuisine all over the world. Ginger and its relatives have also been used extensively in traditional medicine and as the source of dyes and perfumes.

DISTRIBUTION: Possibly native to India but cultivated widely in tropical climates.

USE IN GIN: Ginger is a popular botanical in gin. It imparts a warm, earthy, spicy flavour. It also features prominently in gin cocktails.

# True cinnamon

SCIENTIFIC NAME: *Cinnamomum verum*
FAMILY: Laurel family (Lauraceae)

DESCRIPTION: True cinnamon is a small evergreen tree that grows to about 12 m high. Its glossy leaves have prominent parallel veins and are a bright shade of crimson when young, and turn green with age. The small yellowish-green flowers are silky on the outside, and are followed by brownish-black fruits which are about 1 cm long. This species, and its relatives, are a source of the familiar spice, which is derived from the inner bark.

DISTRIBUTION: Native to Sri Lanka.

USE IN GIN: Cinnamon is often used alongside nutmeg and other spices to give a warm, earthy spiciness to gin. Cassia or Chinese cinnamon is derived from a related species (*Cinnamomum cassia*) and is also used widely in gin formulations.

# Sweet flag

SCIENTIFIC NAME: *Acorus calamus*
FAMILY: Acorus family (Acoraceae)

DESCRIPTION: Sweet flag has numerous other common names including gladdon, cinnamon sedge and sweet root. It is a tall, wetland plant that grows to about 2 m high and has broad, grass-like leaves with crimped edges, and inconspicuous flowers borne in spikes. Powdered rhizome, stembark and leaf pastes derived from the plant have been used in traditional medicine for centuries by many cultures for a host of ailments; the plant was also once thought to be a powerful aphrodisiac.

DISTRIBUTION: Widespread across the Northern Hemisphere.

USE IN GIN: Sweet flag is used occasionally, especially in the formulation of heritage craft gins, or for those with 'foraged' botanicals. The powdered root is vaguely reminiscent of ginger and gives a subtle, spicy earthiness to the gin, with a hint of lemon.

# Liquorice

SCIENTIFIC NAME: *Glycyrrhiza glabra*
FAMILY: Pea and bean family (Fabaceae)

DESCRIPTION: Liquorice is a leafy perennial with purple flowers borne in clusters. It has long been cultivated for its thick underground rhizomes. These are harvested from well-established plants, dried and ground. Although not related to star anise, aniseed or fennel, liquorice yields a similar flavour to these botanicals. It is used in confectionery, often alongside aniseed oil, which is flavoured more powerfully.

DISTRIBUTION: Native to Europe and Asia.

USE IN GIN: This traditional woody botanical has long been used to give gin a distinctive earthy sweetness that is quite unlike the familiar aniseed taste of confectionery containing liquorice, and much more akin to the aroma of wood, cut grass and dried hay.

# Garnishes

All manner of plants are used as garnishes for gin. Citrus fruit, zest and peel are among the most popular garnishes – particularly lemon, lime and grapefruit – as are various herbs including lavender, rosemary and basil. More unusual garnishes include strawberries, mint, cucumber, apple and an assortment of other ingredients which mirror the botanicals used in the formulation, such as star anise.

# References and further reading

1 Ivan Tonutti, 'Wine Aromatisation', in *Potencialidades e Aplicações das Plantas Aromáticas e Medicinais. Curso Teórico-Prático*, ed. A.C. Figueiredo, J.G. Barroso and L.G. Pedro, 3rd edition, Edição Centro de Biotecnologia Vegetal – Faculdade de Ciências da Universidade de Lisboa, Lisbon, 2007, pp.147–54 (p.148).

2 Jessica Warner, 'The naturalization of beer and gin in early modern England', *Contemporary Drug Problems*, 24:2, 1997, pp.373–402 (p.386), https://doi.org/10.1177.009145099702400209 (accessed 10 February, 2019).

3 Ernest L. Abel, 'The gin epidemic: much ado about what?', *Alcohol and Alcoholism*, 36:5, 2001, pp.401–5, https://doi.org/10.1177/009145099702400209 (accessed 10 February, 2019).

4 James Nicholls, 'Gin Lane Revisited: Intoxication and Society in the Gin Epidemic', *Journal for Cultural Research*, 7:2, 2003, pp.125–46 (pp.128–9), https://doi.org/10.1080/14797580305358 (accessed 10 February, 2019).

5 Henry Jeffreys, *Empire of Booze*, Penguin Books, London, 2016, pp.36–7.

6 *The Gin Foundry, Gin Distilled: The essential guide for gin lovers*, Ebury Press, London, 2018, p.4.

7 Ivan Tonutti and Peter Liddle, 'Aromatic plants in alcoholic beverages. A review', *Flavour Fragrance Journal*, Special Issue, 25:5, 2010, pp.341–50 (p.345), https://doi.org/10.1002/ffj.2001 (accessed 6 January 2019).

8 Gin Foundry, *Gin Distilled*, pp.36–7.

For an excellent guide to many of the plants featured here, see David Mabberley, *Mabberley's Plant-book*, Cambridge University Press, Cambridge, 2008.

# Index of beverages

# Index of botanicals

# Index of scientific names

First published in 2020 by the Bodleian Library, Broad Street, Oxford OX1 3BG, in association with Oxford Botanic Garden and Arboretum
www.bodleianshop.co.uk

ISBN: 978 1 85124 553 6

Text © Oxford Botanic Garden and Arboretum, 2020
Illustrations © Chris Thorogood, 2020

Cover design by Dot Little at the Bodleian Library
Designed and typeset by Ocky Murray in Linotype Neue Helvetica
Printed and bound by 1010 International Ltd. on 157gsm Top Kote matt art paper

British Library Catalogue in Publishing Data
A CIP record of this publication is available from the British Library